# YOUR KNOWLEDGE HAS VALUE

Umana Rafiq

# A comprehensive study on properties of Semiconductors and p-n Junction

GRIN Publishing

**Bibliographic information published by the German National Library:**

The German National Library lists this publication in the National Bibliography; detailed bibliographic data are available on the Internet at http://dnb.dnb.de .

**Imprint:**

Copyright © 2012 GRIN Verlag GmbH
Print and binding: Books on Demand GmbH, Norderstedt Germany
ISBN: 978-3-656-71903-8

**This book at GRIN:**

http://www.grin.com/en/e-book/278587/a-comprehensive-study-on-properties-of-semiconductors-and-p-n-junction

**GRIN - Your knowledge has value**

Since its foundation in 1998, GRIN has specialized in publishing academic texts by students, college teachers and other academics as e-book and printed book. The website www.grin.com is an ideal platform for presenting term papers, final papers, scientific essays, dissertations and specialist books.

**Visit us on the internet:**

http://www.grin.com/

http://www.facebook.com/grincom

http://www.twitter.com/grin_com

# Comprehensive Study On Properties of Semiconductors and p-n Junction

Umana Rafiq Ananna

Department of EEE, Ahsanullah University of Science and Technology, Dhaka, Bangladesh.

**Abstract: A comprehensive study of p-n junction is necessary to design an electronic device as well as circuits. An electronic device controls the movement of electrons. The study of electronic devices requires a basic understanding of the relationship between electrons and other components of an atom. This leads to knowledge of the differences between conductors, insulators and semiconductors and to an understanding of p-type and n-type semiconductor material. p-n junction is formed by joining p-type and n-type semiconductor materials. So the concept of semiconductor, majority and minority carrier of p-type and n-type semiconductor, doping, depletion region of p-n junction, mobility and conductivity, drift and diffusion current, carrier concentration calculation and Fermi energy level is actually the comprehensive study of p-n junction.**

## Introduction

Semiconductors as a separate class of materials were known by the end of the 19th century. Not until the development of quantum theory, however, could the characteristics of dielectrics, semiconductors, and metals be understood. But today Semiconductor elements are widely used all over the world. Because of its unique properties it has its popularity in the field of manufacturing electronic devices over semiconductors. However the summery can only describe a small portion of the vast literature regarding semiconductors and the most basic semiconductor device called "p-n junction". So the study of the basic philosophy of a semiconductor device naturally depends on the physics of semiconductors. But at first we have to go through the fundamentals of semiconductors.

## I.  What is a semiconductor?

Semiconductor is a material that behaves in between a conductor and an insulator. At ambient temperature, it conducts electricity more easily than an insulator, but less readily than a conductor. At very low temperatures, pure or intrinsic semiconductors behave like insulators. At higher temperatures though or under light, intrinsic semiconductors can become conductive. The addition of impurities to a pure semiconductor can also increase its conductivity. Examples of semiconductors include chemical elements and compounds such as silicon, germanium, and gallium arsenide.
But before proceeding to the elaborate studies of semiconductors we need to at first understand the properties of solids and their types in which we will find

semiconductor. And it will tell us why semiconductors are different from conductors and insulators.

## II.  Classification of conductor, semiconductor and insulator

On the basis of relative values of electrical conductivity ($\sigma$) or resistivity ($\rho=1/\sigma$), the solids are broadly classified as Metals, Semiconductors and Insulators. They can also be classified on the basis of band theory.
In the metallic conductors conduction band consists free electrons which can be easily moved by the influence of applied electric field. And each time and electrons leaves a hole behind which is then filled up by the electrons from valance band. Electrons from valance band thus move to the conduction band and current flows. There is almost no forbidden gap for the conductors.
Since the Forbidden gap for the insulators is so wide that electrons in the valance band all remain bound and no free electrons are available in the conduction band.
Since a semiconductor has narrow forbidden gap and the valance band is completely full and conduction band is empty. When an external energy is applied some excited valance electrons move into conduction band and thus creates a current flow.

*The orientation of the solid atoms also distinguishes the conductor, semiconductor and insulator.

## III.  Conductivity in Semiconductors

Conductivity depends largely upon what happens to the outer shell electrons when the atoms bond together to form a solid. In the case of semiconductors, they usually have four valance electrons (silicon) and four holes. So when these atoms come close to each other to form a solid piece, the valance electrons behave as if they are orbiting between the valance shells of two atoms. In this way each valance electrons fills one of the holes on the valance shell of neighboring atom which is known as covalent bonding. When semiconductor materials are prepared for manufacturing devices, they are aligned into a three-dimensional crystal lattice where each atom is covalently bonded to 4 surrounding atoms.

The conductivity of semiconductor elements depends on some effects. Semiconductors possess negative temperature coefficient of resistance. Hence their conducting nature increases with rise in temperature. At absolute zero temperature (-273°C) it acts as an insulator.

As the temperature increases, and as the energy gap between the conduction and valence bands is very small (~1eV) so the thermal energy gained by the valence electrons propels them to the conduction band. Because when the temperature of a semiconductor rises above absolute zero, there is more energy in the semiconductor to spend on lattice vibration and on exciting electrons into the conduction band.

When semiconductors are optically excited or excited by the effects of light then it's conductivity increases.The conductivity of an intrinsic semiconductor depends on its temperature, but at room temperature its conductivity is very low. When a small amount maybe few parts per million (ppm), of a suitable impurity is added to the pure semiconductor, the conductivity of the semiconductor is increased manifold. Such materials are known as e extrinsic semiconductors or impurity semiconductors. The deliberate addition of a desirable impurity is called doping and the impurity atoms are called dopant. Such a material is also called a doped semiconductor. The dopant has to be such that it does not distort the original pure semiconductor lattice. It occupies only a very few of the original semiconductor atom sites in the crystal. A necessary condition to attain this is that the sizes of the dopant and the semiconductor atoms should be nearly the same.

Doping has a dependence on Dopant-Site binding energies,

$$E_B \approx - m_n \, X \, q^4 / [2 \, (4 \, X \, \pi \, X \, K_s \, X \, \varepsilon \, X \, \hbar)^2]$$

Where, $E_B$ = Binding energy
$m_n$ = effective mass of charge carrier
$K_s$ = Di-electric constant for silicon

## IV.  Types Of semiconductors

Pure semiconductors are called intrinsic type material. It is the condition of a semiconductor before the process of doping. It cannot be used to make a device. It is free of impurities and crystal Defect. When the temperature increases more thermal energy is available to the electrons and some of them may break away from the valence shell and creates freed negative charges. This ionizing process also creates holes or vacancies in the structure with effective positive charge. But since these charges are randomly distributed in the structure there is no net current flowing through it and it remains in a balanced condition. Apart from the process of generation of conducting charges also occurs a process of recombination in which electrons recombine with holes. And the recombination happens because of the collision of electrons and holes. Extrinsic type materials are semiconductors which can conduct current unlike the intrinsic material. Extrinsic type semiconductor must contain impurity atoms that are added to them by the process of doping. Due to adding impurity atoms (which have their own crystal structure) and creating a spare electron the extrinsic semiconductor has crystal defects.

This kind of semiconductors  contains imbalance in the carrier concentration. The net current flow is not zero.

On the basis of doping Extrinsic semiconductors are of two types:

### A.  n-Type:
n-type semiconductors are made by donor doping. Donor doping generates free electrons in the conduction band. It is done by adding impurity atoms which have to be pentavalent (five electrons and three holes in the valance shell). The semiconductor (tetravalent) atoms have 4 electrons in the valance shell. So 4 of them make a covalent bond with the 4 added impurity electrons. and will produce a spare valance shell electron for each addition. Then each of these spare electrons enters the conduction band and thus improves the conductivity. Since electrons are negative charge carriers so these materials are referred to n-type materials. Here electrons are majority carriers and holes are minority carriers.
For doping tetravalent Si or Ge atom typical donor impurity atoms are:    Arsenic (As), Antimony (Sb), Phosphorous (P) etc.

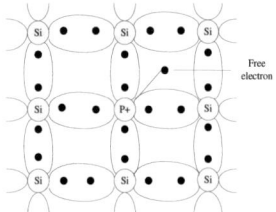

Figure 1: n-type semiconductor

### B.  p-Type:
p-type semiconductors are made by acceptor doping. Acceptor doping produces holes or shortage of electrons in valance band. In this case impurity atoms for doping are trivalent (three electrons and five holes in the valance shell). So during doping the three impurity electrons pair with 3 electrons of semiconductor's valance shell. And thus each of the additions leaves behind a vacancy or holes. And now conduction may occur in the process of hole transfer. These referred to as p-type materials. Typical acceptor impurity atoms are: Gallium (Ga), Boron (B), Aluminium (Al), Indium (In) etc. Here Holes are majority carriers and electrons are minority carriers.

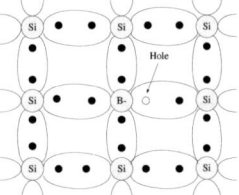

Figure 2: p-type semiconductor

*Neutrality :* By doping of a semiconductor we can only increase the no. of free electrons in the conduction band. or no. of holes in the valance band. But Still it remains electrically neutral. It happens because doping does not change the atomic structure; rather it only changes the no. of charge carriers. This is a kind of recombination which causes the semiconductor to be electrically neutral before applying any external potential.

There are also Semiconductors classified as Direct Band gap and Indirect Band gap Semiconductors. Based On Rated Compositions there are Elementary Semiconductors: Si and Ge. Compound Semiconductors: combinations of elements of Group (iii)-(iv), (iv)-(iv), (iv)-(v). Based On Alloys there are semiconductors known as Binary: GaAs, Tertiary: GeAsP and Quaternary: GaAsInP

## V. Carriers

Semiconductors alike conductors contain charge carrying entities which can create a flow of current. But with breaking the bond between the electron and atom no current flow is possible. Equivalently if we consider the Energy Band Theory then current flow is not possible if the valance band is completely filled with electrons and the conduction band is devoid of electrons. Though in actual case the valance band electrons move about in the crystal but no current arises. It is because The momentum of electrons is quantized in addition to their energy. So even if the valance band is full of electrons the net momentum of the electrons is identically zero and thus no current flow arises.

But when the bond (for example (Si-Si) is broken the released electron can freely wander about the lattice and acts as a negative charge carrier. In terms of the band model, the excited electrons from valance band can move into the conduction band and thus create charge carriers.

In addition, when a bond is broken and electron is released a hole or vacancy is created. Which is immediately replaced by the nearby electron creating another hole and this goes on. In the band model it can be visualized as if a hole that is created in the valance band that is created while an excited electron jumped into the conduction band; is filled by the nearby electron and another hole is created. Thus no. of holes also moves through the vast sea of valance electrons.
These holes are positive charge carriers which can also create a flow of current.

## VI. Difference In Band Structure

The major difference based on which materials are classified is not the nature of the energy band of their band model, rather on the magnitude of the energy gap between the bands. Insulators have wide band gap while semiconductors have a narrow one. In metals the band gap being so small due to the overlap of the conduction and valance band; there is always abundance of charge carriers.

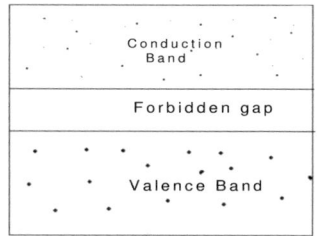

Figure 3: Band Structure

Semiconductors present the intermediate characteristics in between the conductor and insulator. At room temperature (T=300K), $E_G$ = 1.42 eV in GaAs, $E_G$ = 1.12 eV in Si and $E_G$ = .66 eV in Ge. In semiconductors increasing thermal energy can excite the electrons of the valance band into the conduction band creating moderate no. of charge carriers in these materials.

## VII. Carrier Properties

Charge carriers in the doped semiconductor materials have the some unique properties which are very important in the understanding of the characteristics of the semiconductor after doping. The very first property is the generation of Negative charges or electrons and positive charges or holes. The magnitude of the charge of electrons and holes is the same. To three place accuracy in MKS units, $q=1.60X10^{-19}$Coulomb.

Mass, like charge , is another very basic property of carriers. The apparent or effective mass of electrons within a crystal is a function of the semiconductor material (Si or Ge) and is significant from the mass of electrons within a vacuum. It allows us to conceive of the electrons and holes as Quasi-classical particles and to employ classical particle relationships in most device analysis. Effective masses can have multiple components. It also varies with temperature. If electrons effective mass is $m_n$, the force on the charge is,

$$F=-qXE=m_nX(dv/dt)$$

and for the holes,

$$F=qXE=m_pX(dv/dt)$$

Another significant property is the concentration of carriers. Which is different in intrinsic and extrinsic semiconductors.

## VIII. State and Carrier Distribution

In the study of p-type and n-type semiconductors, it is very important to calculate the precise numerical value of the of the carrier concentrations in them. Another property that is needed to be understood is the distribution of carriers as a function of energy in the respective energy bands and also te carrier concentration in the semiconductors under equilibrium conditions.

## A. Density of States:

The state distribution is an important component to determine carrier distributions and concentrations. To determine the density of states, an analysis need to be performed on the basis of quantum mechanical considerations,

$$g_C(E) = \frac{m_n\sqrt{2m_n}X(E-E_C)}{\pi^2\hbar^3} \qquad E > Ec$$

$$g_v(E) = \frac{m_n\sqrt{2m_p}X(E_V-E)}{\pi^2\hbar^3} \qquad E <= Ev$$

Where, $g_C$ (E)=density of states at an energy E in conduction band.
$g_C$ (E) = density of states at an energy E in valance band
$m_n$ = effective mass of electrons
$m_p$ = effective mass of holes.

## B. The Fermi Function :

The Fermi Function $f(E)$ specifies, under equilibrium conditions, the probability that an available state at an energy E will be occupied by an electron. Mathematically, the Fermi function is simply a probability distribution function.
In mathematical symbols,

$$f(E) = 1/(1 + e^{(E-E_F)/kt})$$

where,
$E_F$ = Fermi energy or Fermi level
$k$ = Boltzmann constant (K = 8.617 )
$T$ = temperature in Kelvin(K)

The Fermi function  is universal in the sense that it applies with equal validity to all materials – insulators, semiconductors and metals. Finally, the relative positioning of the Fermi energy $E_F$ compared to $E_c$ (or $E_v$),an item of obvious concern, is treated in subsequent subsections. It is a temperature dependent function.

## C. Equilibrium Carrier Concentrations :

We have arrived at an important point in the carrier modeling process, For the most part, this section simply embodies the culmination of our modeling efforts, with working relationships for the equilibrium carrier concentration being established to complement the qualitative carrier information presented in previous sections.

## D. Formulas for n and p :

Integration over the equilibrium distribution of electronics in the conduction band yields
The equilibrium electron concentration. A similar statement can be made relative to the hole concentration. We therefore conclude

$$n = \int g_c(E) f(E) dE$$

$$p = \int g_v(E)[1 - f(E)]dE$$

Identifying $F_{1/2}(\eta) = \eta^{1/2} d\eta / 1 + e^{(\eta - \eta c)}$, the Fermi-Direc integral of order ½ one obtains

$$n = N_c X e^{(E_F - E_c)}/kT$$

$$p = N_v X e^{(E_v - E_F)}/kT$$

## E. The $n_i$ and $np$ Relation:

The intrinsic carrier concentration can figure prominently in the quantitative calculation of the carrier concentrations. Continuing to establish pertinent carrier concentration relationships, we next interject considerations specifically involving this important material parameter.
Firstly, one obtains

$$N_i = \sqrt{(N_c N_v)} X e^{-Eg/2kT}$$

A second very important $n_i$-based relationship is.

$$np = n^2_i$$

This relationship often proves to be extremely useful in practical computations.

# IX.   Carrier Concentration of Electrons

The equation for the thermal equilibrium concentration of electron can be found by integrating over the conduction band energy or,

$$n_0 = \int g_C(E) f_F(E) dE$$

The lower limit of integration is $E_C$ and upper limit of integration should be top of the conduction band energy. Since the Fermi probability function rapidly approaches zero with increasing energy we can  take the upper limit of the integration to be infinity.

We are assuming that the Fermi energy is within the forbidden energy band-gap. For electrons in the conduction band we have  the Fermi probability function reduced to Boltzmann approximation,

$$f_F(E) = 1/[1+ e^{E-EF/kT}] \approx e^{E-EF/kT}$$

So, the thermal equilibrium density of electrons in the conduction band is found from,

$$n_0 = 4\pi (2mnkT/h^3)^{3/2} \cdot e^{\wedge-EC-EF/kT} \int_0 \eta^{1/2} e^{\eta} \, d\eta$$

The integral is the gamma function with a value of,

$$\int_0 \eta^{1/2} e^{\eta} \, d\eta = \frac{1}{2} \sqrt{\pi}$$

Then the value of $n_0$ will be after simplification,

$$n_0 = N_C \cdot e^{-EC-EF/kT}$$

# X.   Carrier Concentrations of Holes

The thermal equilibrium concentration of holes in the valance band can be found by the equation of the distribution of holes over the valance band energy,

$$p_O = \int g_V(E)[1 - f_F(E)] \, dE$$

For energy state in the valance band, $E < E_V$. If $(E_F - E_V) \gg kT$ then the Boltzmann approximation becomes,

$$f_F(E) = 1/[1 + e^{E-EF/kT}] \approx e^{E-EF/kT}$$

so further simplification yields,

$$n_O = N_V \cdot e^{-EF-EV/kT}$$

## XI. Position of Fermi energy level

The thermal equilibrium concentration of holes in the conduction band is,

$$p = N_v \, e^{-(E_F - E_V)/KT}$$

where $N_v$ is the effective density states in the valance band. The thermal equilibrium concentration of electrons in the conduction band is,

$$n = Nc \, e^{-(E_C - E_F)/KT}$$

So, the Fermi energy level is,

$$E_F = \frac{1}{2}\left\{(E_V + E_C) + KT \ln \frac{nN_V}{pN_C}\right\}$$

### A. For Intrinsic Semiconductor:
Where the number of electrons are equal to the number of holes and the effective density of states of valance band we can write the equation of Fermi energy level,

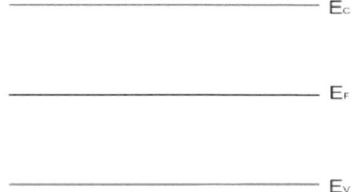

Figure 4: Ferni level position in Intrinsic material.
We can see that, the Fermi energy level is just in the middle of the conduction band and valance band. So there will be no flow of current.

### B. For Extrinsic semiconductor:
In n-type material as the electron concentration is higher, the Fermi Level moves closer to the conduction band.

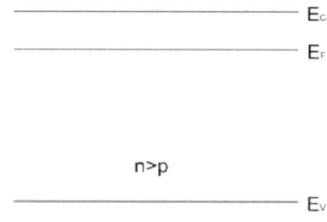

Figure 5: Fermi level osition in n-type materal

In p type material, as the holes concentration is higher the Fermi level moves closer to the valance band.

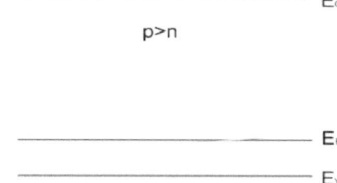

Figure 6: Fermi level position in p-type material.

## XII. Basic Structure of p-n junction

When a p-type semiconductor is joined to an n-type semiconductor such that the crystal structure remains continuous at the boundary, the junction is called p-n junction and the resulting combination is called p-n junction diode or semiconductor diode. Figure 7 refers to the p-n junction and figure 8 refers to the diffusion that occurs in p-n semiconductor.

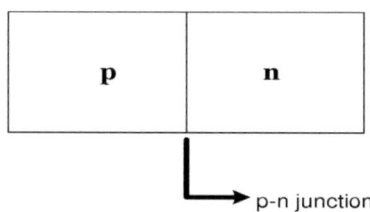

Figure 7: Simplified geometry of a p-n junction

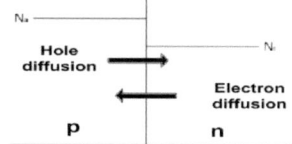

Figure 8: Doping profile of an ideal uniformly doped p-n junction.

In a p-n junction, the doping concentration is uniform in each region but there is an abrupt change in doping at the junction. We know, when there is concentration gradient, there is diffusion .So the majority carrier in p region will diffuse towards n-region will diffuse towards p-region. But, diffusion will not be for infinite amount of time. When an electric field is created the holes in p-type will be directed in the same direction as the electric field and the electrons in n-type will be directed just in opposite. So, in order to create an electric field, we need to bias the conductor.

## XIII.   Depletion region

When a p-type and an n-type conductor are joined, the junction contains some uncovered positive charge and uncovered negative charge. This region is called depletion region. When the conductor is biased to get an electric field, then due to flow of electrons and holes in the opposite direction, gradually the depletion region decreases. In figure 10 we can see the depletion region that occurs in p-n junction.

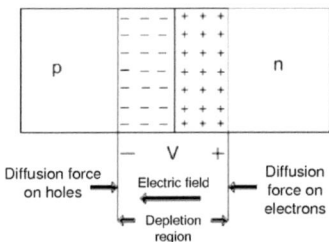

Diffusion force on holes →
Electric field
Diffusion force on electrons
← Depletion region →

Figure 9:  The depletion region, the electric field and the forces acting on the charged carriers.

## XIV.   Built in potential barrier

When voltage is applied at the p-n junction then the junction is in thermal equilibrium. Due to absence of any charge carrier at the junction a potential difference is created at the junction. This potential difference creates a barrier which prevents the electrons of n-type to move to p-type or vice versa. This potential barrier is known as built in potential barrier.

We can determine the distance that space charge region extends into the p and n regions from the metallurgical junction. This distance is known as the space charge width. We may write, for example

$$X_p = \frac{NdXn}{Na}$$

Then the equation of the space charge width of the depletion region, $x_n$ is,

$$X_n = \{\frac{2 \epsilon sVbi}{e}[\frac{Na}{Nd}]/\frac{1}{Na+Nd}\}^{1/2}$$

The above equation gives the width of the depletion region, $x_n$ extending into the n-type region for the case of zero applied voltage.
Similarly, $x_p$ becomes-

$$X_p = \{\frac{2 \epsilon sVbi}{e}[\frac{Nd}{Na}]/\frac{1}{Na+Nd}\}^{1/2}$$

Where, $X_p$ is the depletion region extending into the p-region for the case of zero applied voltage. The total depletion width W is the sum of the two components,

$$W = X_n + X_p$$

$$W = \{\frac{2 \epsilon sVbi}{e}[\frac{Na+Nd}{Na.Nd}]\}^{1/2}$$

## XV.   Forward Bias

A forward bias is established by applying the positive potential to the p-type material and negative potential to the n-type material. In figure 11 forward bias occurs as diffusion stops due to application of electric field.

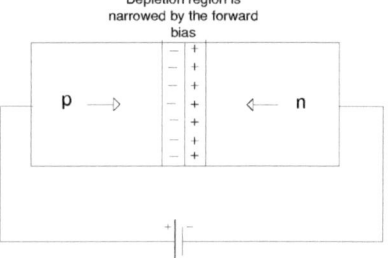

Depletion region is narrowed by the forward bias

Figure 10: Forward biasing p-n junction

Due to forward bias the electrons and the holes will move in the opposite direction gradually decreasing the width of the depletion region. Initially, small amount of current flows due to the presence of minority carriers. After a certain time when there is no depletion region huge amount of current flows through the conductor.

## XVI.   Reverse Bias

If an external potential is applied across the p-n junction such that the positive terminal is connected to the n-type material and the negative terminal is connected to the p-type, then it is known as reverse bias. In figure 12 we see reverse bias occurs and the depletion region increases.

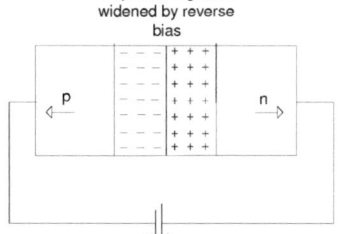

Deplation region is
widened by reverse
bias

p n

Figure 11: Reverse Biasing p-n junction

Under reverse bias the depletion region widens .As a
result, the electric field produced by the ions cancel out
the applied reverse bias.

## XVII.  p-n Junction Current

When a forward-bias voltage is applied to a p-n junction, a
current will be induced in the device. We initially consider
a qualitative discussion of how charges flow in the p-n
junction and then consider the mathematical derivation of
the current-voltage relationship.

### A.  Ideal Current-Voltage Relationship
The ideal current-voltage relationship of a p-n junction is
derived on the basis of our assumptions.

1. The abrupt depletion layer approximation applies. The
space charge regions have abrupt boundaries and the
semiconductor is neutral outside of the depletion region.
2. The Maxwell-Boltzmann approximation applies to
carrier statistics.
3. The concept of low injection applies.
4a. The total current is a constant throughout the entire pn
structure.
4b. The individual electron and hole currents are continues
functions through the p-n structure.
4c. The individual electron and hole currents constant
throughout the depletion region.

### B.  Boundary Condition:
The n region contains many more electrons in the
conduction band than the p region, the built-in potential
barrier prevents this large density of electrons from
flowing into the p region. The built-in potential barrier
maintain equilibrium between the carrier distributions on
either side of the junction.
An expression for the built-in potential barrier is

$$V_{bi} = V_t \ln\left(\frac{NaNd}{ni^2}\right)$$

If we divide the equation by $V_t=kT/e$, taking the
exponential of both sides, and then taking the reciprocal,
we get,

$$\frac{ni^2}{NaNd} = exp\left(\frac{-eVbi}{kT}\right)$$

From the above equation, we can write,

$$n_{po} = n_{no}exp\left(\frac{-eVbi}{kT}\right)$$

This equation relates the minority carrier concentration on
the p side of the junction to the majority carrier electron
concentration on the side n of the junction in thermal
equilibrium.
If a positive voltage is applied to the p region with respect
to the n region, the potential barrier is reduced. The
electric field $E_p$ induced by the applied voltage is in the
opposite direction to the thermal equilibrium space charge
electron field, so the net electric field in the space charge
region is reduced below the equilibrium value. As long as
the bias $V_a$ applied, the junction of carriers across the
space charge region continues and a current is created in
the p-n junction. This bias condition is known as forward
bias. The energy-band diagram of the forward-biased p-n
junction.
The potential barrier $V_{bi}$ n the above equation can be
replaced by $(V_{bi}-V_a)$ when the junction is in the forward
biased. Then the equation becomes-

$$n_p = n_{no}exp\left(\frac{-e(Vbi-Va)}{kT}\right) = n_{no} exp\left(\frac{-eVbi}{kT}\right) exp\left(\frac{-eVa}{kT}\right)$$

If we assume a low injection. the majority carrier electron
concentration $n_{n0}$ does not change significantly. However,
the minority carrier concentration $n_p$, can be deviate from
its thermal –equilibrium value $n_{p0}$ by orders of magnitude.

$$n_p = n_{p0} exp\left(\frac{eVa}{kT}\right)$$

Exactly, the same process occurs for majority carrier holes
in the p region which are injected across the space charge
region into the n region under a forward-bias voltage, we
can write that,

$$p_n = p_{n0} exp\left(\frac{eVa}{kT}\right)$$

Where $p_n$ is the concentration of minority carrier holes at
the edge of the space charge region in the n region. The
above Fig. shows these results. By applying a forward-bias
voltage, we create excess minority carriers in each region
of the pn junction.

### C.  Ideal p-n Junction Current:
The total current in the p-n junction is the sum of the
individual electron and hole currents which are constant
through the depletion region. Since the electron and hole
currents are continues functions through the p-n junction
,the total p-n junction current will be the minority carrier
hole diffusion current at $x = x_n$ plus the minority carrier
electron diffusion current at $x = -x_p$. The gradients in the
minority carrier concentration produce diffusion currents
and since we are assuming the electric field to be zero at
the space charge edges, we can neglect any minority
carrier drift current component.

We can calculate the minority carrier hole diffusion current density at $x=x_n$ from the relation

$$J_p(x_n) = -eD_p \frac{dPn(x)}{dx}\Big|_{x=xn}$$

Since we are assuming uniformly doped regions, The thermal-equilibrium carrier concentration is constant, so the hole diffusion current density may be written as,

$$J_p(x_n) = -eD_p \frac{d(\wp Pn(x))}{dx}\Big|_{X=Xn}$$

$$Or, \ J_p(x_n) = e\,Dp\,\frac{eDpPno}{Lp}\,[exp(\frac{eVa}{kT}) - 1]$$

The hole current density for this forward bias condition is in the +direction, which is from the p to the n region. Similarly, we may calculate the electron diffusion current density at $x= -x_p$. This may be written as,

$$J_n(-x_p) = -eD_n \frac{d(\wp Np(x))}{dx}\Big|_{X=-Xp}$$

Using above equation, we obtain,

$$J_n(x_p) = e\,Dn\,\frac{eDnNpo}{Ln}\,[exp(\frac{eVa}{kT}) - 1]$$

The electron current density is also in the +x direction. The total current density in the p-n junction is then,

$$J= J_p(x_n)+ J_n(x_p)=[\frac{eDpPno}{Lp} + \frac{eDnNpo}{Ln}][exp(\frac{eVa}{kT}) - 1]$$

The above equation is the ideal current-voltage relationship of a p-n junction. We may define a parameter $J_s$ as,

$$J_s =[\frac{eDpPno}{Lp} + \frac{eDnNpo}{Ln}]$$

So that the above equation may be written as,

$$J = J_s[exp(\frac{eVa}{kT}) - 1]$$

The above equation is known as the ideal-diode equation, gives a good description of the current-voltage characteristics of the p-n junction over a wide range of currents and voltages.

## Conclusion

In conclusion, without p-n junction the world of electronics cannot be imagined. In fact, p-n junction has brought a revolutionary change in the field of electronics. After having a comparative study of p-n junction, we are able to understand the characteristics of diode which is now used in nearly every aspect of electronics. That is why the comparative study of p-n junction is mandatory.

## References

[1] David A. Bell, "Electronic Devices and Circuits", Fifth Edition, US, 2009.

[2] Donald A. Neamen, "SEMICONDUCTOR PHYSICS AND DEVICES", Third Edition, New York, 2002.

[3] Ben G. Streetman & Sanjay Kumar Banerjee, "SOLID STATE ELECTRONIC DEVICES", Sixth Edition, USA.

[4] Lectures on Electronics-I by Mohammad Ziaur Rahman, Ahsanullah University of Science and Technology, 2012.